A DISCOVERING SCIENCE BOOK

MAKE A SPLASH!

Science Activities with Liquids

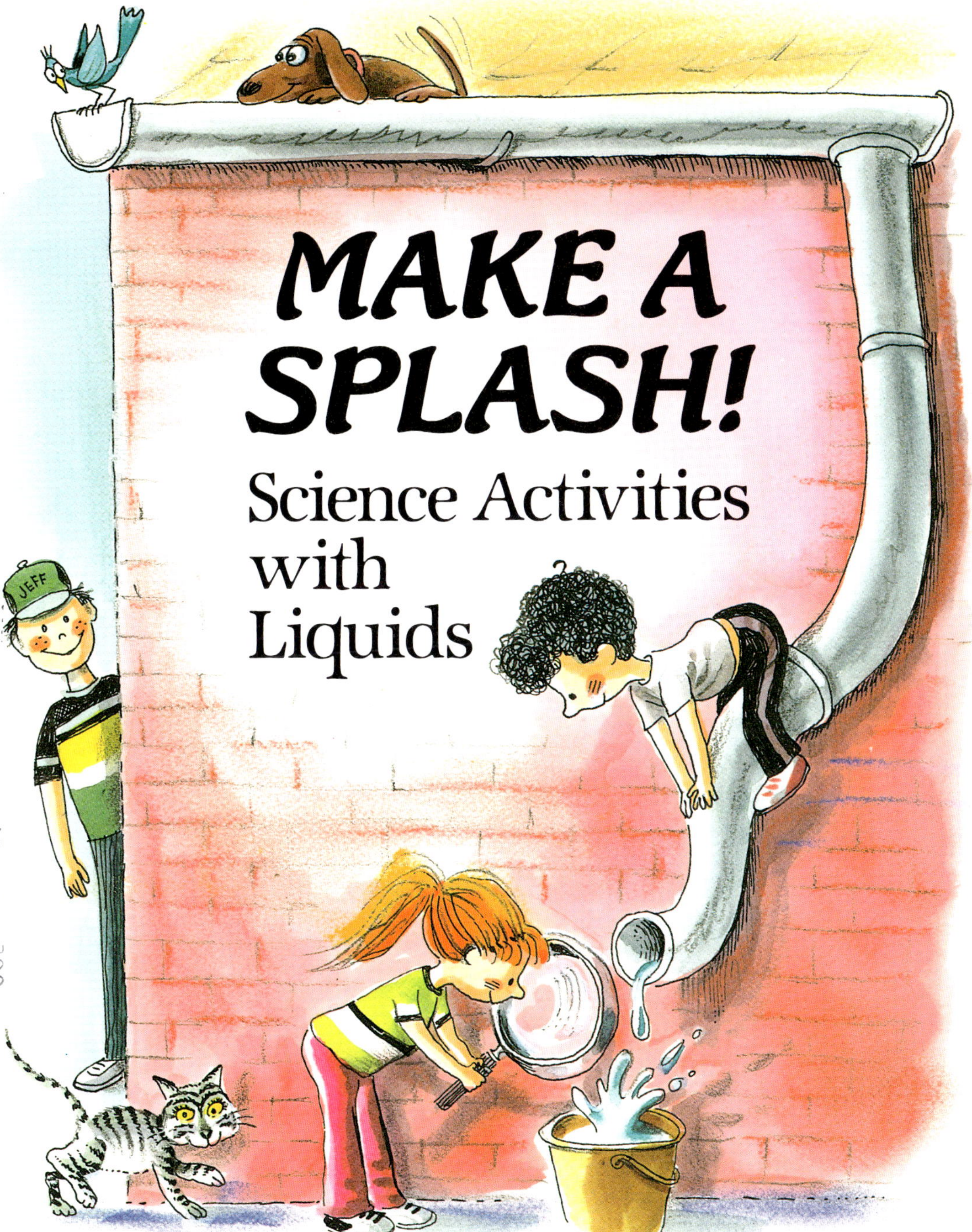

by **Penny Raife Durant**

Illustrated by **John Huehnergarth**

Franklin Watts
New York London Toronto Sydney
1991

This book is dedicated
with love to my son Adam,
whose curiosity is a
continual inspiration to me.
P.R.D.

To Martha, who
raised three children,
one of them a cartoonist.
J.H.

The detergent brands recommended by the author on page 14 are used with the permission of The Procter & Gamble Company.

Library of Congress Cataloging-in-Publication Data
Durant, Penny Raife.
Make a splash: science activities with liquids/by Penny Raife Durant: illustrated by John Huehnergarth.
p. cm.—(A Discovering Science book)
Summary: Demonstrates the physical properties of water and other liquids through simple experiments.
ISBN 0-531-10971-2
1. Liquids—Experiments—Juvenile literature. [1. Liquids--Experiments. 2. Experiments.] I. Huehnergarth, John, ill. II. Title. III. Series. 90-34063
QC145.24.D87 1991 CIP
530.4'2—dc20 AC

Design by Barbara Powderly

Printed in the United States of America
6 5 4 3 2 1

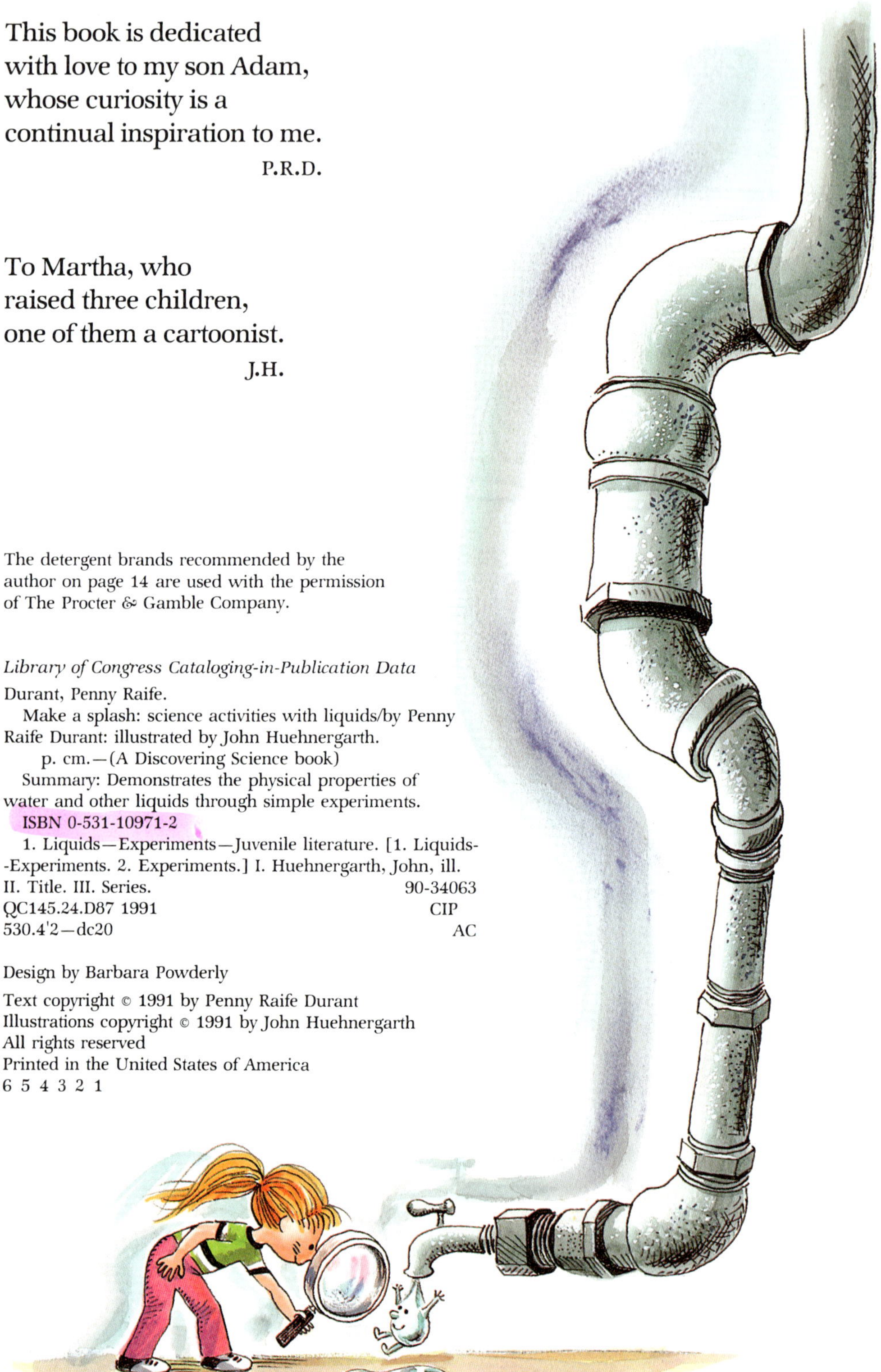

Contents

LEMON

Hi! My name's Mark. Mom says it should be Question Mark, because I ask so many questions. I'm just curious. I'm like my older brother Jeff. I guess Rufus, my dog, is curious, too. He always sticks his nose in things.

My best friend is Sandy. She likes to check into things. She wants to be a detective, like Sherlock Holmes. She always carries a *magnifying glass* with her. It's great for looking at small objects.

In this book, you're going to learn about *liquids*. It'll be fun to read, but don't just read it. Try the activities for yourself. Scientists and detectives are very careful when they work. You need to be careful, too.

Check with an adult first. An adult can help you get all the things you'll need. The adult will do the activities with you. But don't let the adult do them *for* you! These are for you, so you can find out about liquids, too.

WHAT ABSORBS WATER?

I spilled some water on the side of the sink. It beaded up in a puddle. I leaned against the spill. The water soaked into my shirt.

I wondered what else would soak up water. Mom said the word is *absorb*. If the material soaks up water, it absorbs. If water beads up like it did on the sink, the material *repels* water.

I used an *eyedropper* to test lots of things. I made a list of things that absorbed or repelled water. I wasn't sure on which list to put Rufus. He licked off the water every time.

What you need:

glass of water
eyedropper
two lists:
  Absorbs Water
  Repels Water
pencil
items to try, such as:
  cotton ball
  piece of waxed paper
  piece of aluminum foil
  newspaper
  scrap of cloth
  coffee filter

What you do:

1. Guess which items will absorb water. Guess which items will repel water.
2. Test your guesses by dropping two or three drops of water on each item.
3. Did the water soak in? Write the object's name on your *Absorbs Water* list.
4. Did the water bead up on the item? If so, write it on your *Repels Water* list.
5. Try other things. First ask an adult if it's okay.

DO ALL LIQUIDS MIX?

My friend Sandy came over. I was checking to see if waxed paper absorbs or repels water. "Look at it through a magnifying glass," she said.

The water beaded up into funny fat shapes. They were rounded on the sides. We tried to cut the drops in half. Then we blew the drops to make them move.

I pushed drops together to make one big drop. Sandy made twenty little ones.

What you need:

eyedropper
jar of water
pieces of waxed paper
magnifying glass
toothpicks
straws

What you do:

1. Use the eyedropper. Put drops of water on the waxed paper.
2. Look at them through the magnifying glass.
3. With a toothpick cut a drop in half. Can you do it? Can you push two drops together?
4. Using a straw, blow the drops. Can you blow them together?

Other things you might try:

Add a little food coloring to the water. Can you see the shapes better?
On waxed paper, put three drops of cooking oil. Add three drops of food coloring. Try to mix them with a toothpick. What happened?

WHAT FLOATS? WHAT SINKS?

One day Jeff dropped pennies into a glass of water. The pennies sank to the bottom of the glass. Pennies *sink*. Yet the ice in Dad's lemonade always stays at the top. It *floats*.

Sandy looked at the ice with her magnifying glass. "The ice floats," she said.

"What else floats?" I asked. We tried a lot of things. We asked Mom before we tried things. We didn't want to ruin anything. We made two lists: *Things That Float* and *Things That Sink*.

What you need:

sink or dishpan of water
two lists:
  Things That Float
  Things That Sink
pencil
items to try, such as:
  cork
  metal spoon
  key
  wooden block
  piece of aluminum foil

What you do:

1. Fill the sink or dishpan with water.
2. Try to guess about each item. Will it float? Will it sink?
3. Drop each item one at a time in the water. Did it float? Write it on your *Things That Float* list. Did it sink? Write it on your *Things That Sink* list.
4. Are there more things that float? Are there more things that sink?

CAN YOU MAKE IT SINK?

Sandy made a ball of aluminum foil. She tested it. It floated. We flattened it out. It didn't sink.

"I can sink it," Jeff said. He put a heavy metal pan on it. The foil sank.

"That was heavy," Sandy said. "Could it hold something lighter?"

I got my jar of marbles. We folded up the sides of the foil. That way the marbles wouldn't roll off. Sandy made a boat. I made a boat. Hers held more marbles.

What you need:

sink or dishpan of water
aluminum foil, about 12 inches square
marbles

What to do:

1. Fold the foil into the shape of a boat. Give it sides so that the marbles won't roll off.
2. Add marbles, one at a time, until the boat sinks. How many marbles did it hold before it sank?
3. Make another shape from the foil.
4. Add marbles until it sinks. How many marbles did this boat hold?

Other things you might try:

1. Put a Styrofoam meat tray in the water. Can you add enough marbles to sink it? What else might sink it?

CAN YOU MAKE BUBBLES?

Rufus barks at anything that flies. He likes to chase birds and butterflies. He never caught anything until the day Sandy came over to blow bubbles.

We had made bubble *solution* the day before. We poured it into a dishpan. We tried many things to make bubbles. Some things made lots of little bubbles. Some made just one big bubble. Rufus chased them all. He caught some.

What you need:

Bubble solution:

½ cup liquid detergent (Dawn® or Joy® detergents work best)

2 tsp glycerine

½ gallon water

Let sit overnight. This solution can be made in an empty gallon milk jug. Don't try to save it in the plastic jug for long. The jug will leak after some time.

Bubble makers:

pipe cleaners

plastic six-pack beverage holder

funnel

bottom from a strawberry basket

plastic macrame hoops of all sizes

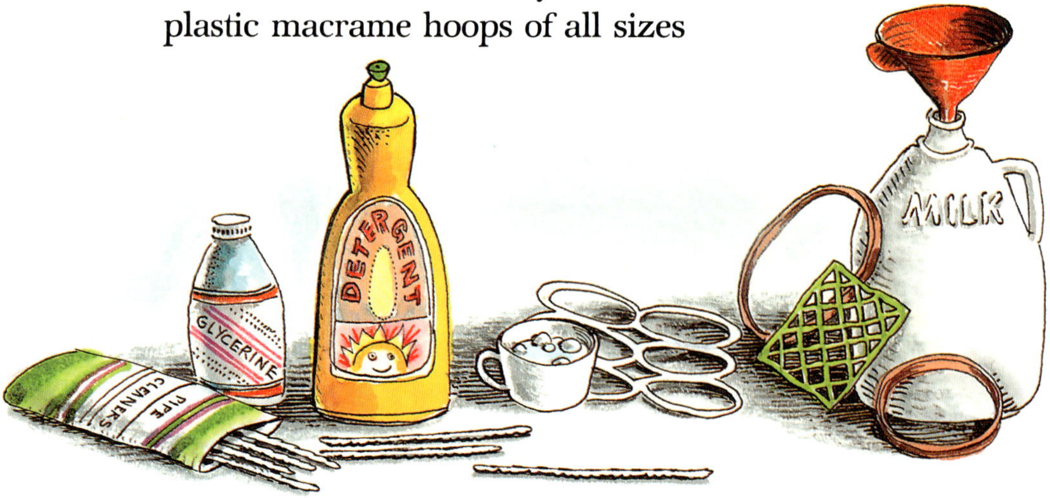

What you do:

1. Pour the solution into a cake pan or dishpan.
2. Make shapes with pipe cleaners.
3. Dip the bubble makers into the solution. Get a lot of solution on your bubble maker. Wave the bubble maker gently through the air.
4. Can you make lots of little bubbles? Can you make a big one? Can you make a square bubble?

IS A LIQUID ALWAYS A LIQUID?

On a hot day, Sandy and I like a frozen treat. One day I took a good look at the drips. "They look like milk," I said.

Sandy peered through her magnifying glass. "They are liquid, like milk. But the treat on the stick is hard. It's *solid*, like ice. Could the drips be frozen again? Could they become solid?"

There weren't enough drips to try it. Rufus licked them all. Mom helped us make frozen treats. You can make some, too. You shouldn't put most science activities in your mouth, but this is one you can *eat*!

What you need:

juice
2 small paper cups
spoon or craft stick for handle

What to do:

1. Mix the juice well.
2. Fill the cups about two-thirds full of juice. Add spoon or stick to each.
3. Put cups in freezer. Be careful not to spill.
4. Check after about an hour. When you can't pull out the stick, it's ready to eat. Tear paper cup off your frozen treat. Look at it through a magnifying glass, then eat and enjoy!

PEAS
Huehnergarth

WHAT HAPPENS TO LIQUIDS IN THE SUNSHINE?

I dropped a piece of my frozen treat on the sidewalk. Rufus licked it up. He left liquid on the sidewalk where he licked. After a few seconds, it was gone. What happened to it? I dipped a stick into water. I used the stick like a pen. The water was my ink. I wrote my name on the warm sidewalk. It stayed there for just a minute. Then it *disappeared.*

Sandy looked with her magnifying glass. "It's gone," she said. "Let me try." She wrote her name with water. It faded away, too. "This is certainly a mystery," she said.

"The water *evaporated,*" Jeff said. "It's now part of the air."

What you need:

paper or plastic cup
water
stick or paintbrush

What you do:

1. On a warm day, dip the stick into water. Write or draw on a cement sidewalk in the sunshine.
2. Wait and watch.
3. What happened to the water? It evaporated.

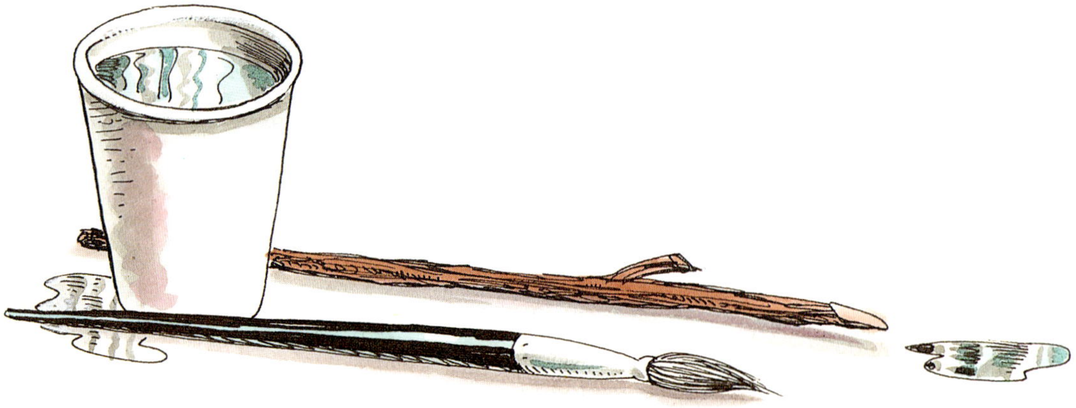

Mark

WHAT IS LEFT WHEN LIQUIDS EVAPORATE?

We had to read fast before the words evaporated.

"Detectives need to write secret messages," Sandy said. "Water evaporates too quickly. We need disappearing ink."

We looked in her detective book. We found a recipe for disappearing ink. Now we can write secret messages to each other. If Jeff finds a message, he can't read it!

What you need:

cotton swab
lemon juice (bottled or fresh)
plain white paper
lamp

What you do:

1. Use the cotton swab as a pen. Use the lemon juice as ink. Write a message on plain white paper.
2. Let the water in the lemon juice evaporate.
3. Hold the paper next to a lighted lamp bulb. Move the paper back and forth until the writing appears. Don't touch the light bulb. It will be hot.

Other things you might try:

1. Use other liquids to make disappearing ink. Try milk, apple juice, yogurt, sugar water. Which ones work?
2. Have an adult help you. Try ironing your secret message on a low setting. Don't use an iron by yourself!

WHAT WILL DISSOLVE IN WATER?

Sometimes I help Mom fix dinner. "Add a teaspoon of salt to the water in this pan," she said.

I measured and poured it in. It disappeared! "What happened?" I asked.

"It *dissolved*," Mom said. "It mixed with the water to make a solution."

Does anything else dissolve in water and make a solution? I wondered.

What you need:

small bowl of water
spoons, for stirring and scooping
things to try:

- sugar
- cinnamon
- cooking oil
- brown sugar
- food coloring

What you do:

1. Think about each item. Do you think it will dissolve in water?
2. Start with fresh water each time.
3. Stir a small amount of each item into the water.
4. Watch to see if it dissolves. Check your guesses.

Other things you might try:

1. Add one spoon of salt to one-fourth cup water. Stir until it dissolves. Leave the solution uncovered until the water evaporates. What happened?
2. Have an adult help you. Boil a cup of water. Keep adding sugar until no more sugar crystals will dissolve. Tie a string on a pencil. Tie a clean button to the other end of the string. *Suspend* the string in a glass. Pour the sugar water into the glass. Let it sit for several days. What happens? You shouldn't taste most crystals, but these you can eat.

CAN YOU MAKE COLORS?

I tried to dissolve food coloring. My favorite color is orange. There's no orange food coloring. I decided to *experiment*.

I added two drops of red and two drops of green to a jar of water. I stirred it. Wow! I got a whole new color! But not orange.

I used fresh water and mixed drops until I got orange. Can you make the color orange? Or the color purple?

What you need:

water
food coloring:
  green
  red
  blue
  yellow
spoon
clean glasses or jars
sheet of paper for recording your answers
pencil or crayons
toothpick

What you do:

1. Fill a jar half-full of water. Drop one or two drops of coloring into it. Record what you do as you experiment.
2. Add another color. Stir. What new color did you make? Add a third color. What color is it now?
3. Throw out the water. Use clean water each time. Be sure to record what you do.

Other things you might try:

Fill a jar half-full of water. Let it sit overnight without moving it. Next day add one drop of coloring. Don't stir. Watch what happens.

CAN YOU MIX COLORS AND KEEP THEM?

Sandy made purple water. She kept the jar on her dresser. Someone knocked it over. She used her magnifying glass. She followed the purple footprints to her cat's bed.

She washed them off the floor. "I wish we could make purple and keep it," she said.

"Hmmm," I said. "Could we *dye* something purple and let it dry? Then we could keep it."

What you need:

old margarine tubs
water
food coloring
coffee filters
newspaper

What you do:

1. Fill three tubs half-full of water. Add a generous amount of coloring. This is dye. Be careful. Don't let the dye touch anything but newspapers.
2. Fold a coffee filter in half. Fold it in half again. Fold it in half again.
3. Dip each corner of the filter in a different color of dye.
4. Open the filter. Let it dry flat on newspapers.

Other things you might try:

1. Try different folds. Can you make a bull's eye?
2. Try dyeing a piece of white yarn. Make a bookmark with it.

FOOD COLORING
1
FOLD
2
FOLD
3
FOLD
4
5
6

CAN WATER MOVE COLORS?

One day I spilled a little water on Jeff's notebook. Was he mad! "The water will evaporate," I said. "The paper will dry."

"Look at the ink!" he yelled. The ink had dissolved in the water. Blue puddles stood where words had been. The blue lines had moved to the edge of the water spots.

"Uh-oh," I said. "Sorry." I left in a hurry. Why had the water taken the ink with it? Can water move marker ink, too?

What you need:

watercolor markers
coffee filters, cut in 1-inch strips
water
glass or jar

What you do:

1. Draw a thick line across a coffee filter strip, two inches from the end.
2. Add an inch of water to the jar.
3. Drape the filter over the side of the jar. Put the bottom of the filter in the water. The marker line should be above the water.
4. Let it sit for about ten minutes.
5. What happened to the marker line?
6. Try markers of different colors.

Other things you might try:

1. Try using a *permanent marker*. Will water move the ink?
2. Try using notebook paper. Will the water move the marker as quickly?

1
2
3
WATERCOLOR
INCHES

Sandy and I keep finding things to try with liquids. I hope you will, too. Whenever you have a question, ask it! Adults don't always know the answer. But they can help you find the answer. They can help you experiment so you can find out for yourself.

Rufus's favorite experiment was the frozen treats. He loves to eat. Or was it the bubbles? He loves chasing things that fly.

Oh! Someone just slipped a message under my door. I think it's from Sandy. I can't read it. The paper is blank. Maybe she used invisible ink. I'd better hold it over a light bulb. It might be important. Good-bye!

GLOSSARY

Absorb (ab-SORB). Take in or soak up. *The sponge absorbs water.*
Disappear (DIS-uh-PEER). No longer be seen. *The sun will disappear behind the clouds.*
Dissolve (di-ZOLV). Become part of a liquid. *Salt will dissolve in hot water.*
Dye (die). Color or stain. *I want to dye my white coat red.*
Evaporate (ee-VAP-uh-RATE). Become part of the air. *The water in an uncovered pan will evaporate.*
Experiment (ek-SPER-uh-ment). Test or try to find out something. *I will experiment to see what happens.*
Eyedropper (I DROP-r). Hollow tube with a rubber squeeze top and tapered end. *Use the eyedropper to add one drop of water.*
Float (flote). Stay on top of water. *The cork will float.*
Liquid (LIK-wid). Something that flows like water. *The ice melted to form water, a liquid.*
Magnifying glass (MAG-nuh-FI-ing glas). Curved glass that makes things look larger. *The ants looked big through the magnifying glass.*
Permanent marker (PER-muh-nent MAR-ker). Marker with ink that is intended to last and not be washed away. *The permanent marker got on my skirt and won't wash out.*
Repel (rih-PEL). Keep off or out; fail to mix with. *My raincoat will repel water.*
Sink (singk). Go underwater. *The boat will sink because it has a big hole in it.*
Solid (SOL-id). Not liquid or gas, firm or hard. *A table is solid, but water is a liquid.*
Solution (so-LU-shun). Liquid formed by dissolving. *When you stir salt into hot water, you get a solution.*
Suspend (suh-SPEND). Hang from above. *The light is suspended from the ceiling.*

About the Author

Penny Raife Durant is a former elementary school teacher with a B.S. in elementary education and an M.A. in early-childhood education from the University of New Mexico. She is co-founder and director of the Calico Butterfly Preschool in Albuquerque, New Mexico, where she lives with her husband and their two sons.

About the Artist

John Huehnergarth was graduated from the Philadelphia Museum College of Art, and has been a free-lance artist for thirty years. His work has appeared in various children's magazines and on over twenty covers of *Newsweek*. He lives with his wife in Princeton, New Jersey.